STARFISH

by Emma Bassier

Cody Koala

An Imprint of Pop!
popbooksonline.com

abdobooks.com
Published by Pop!, a division of ABDO, PO Box 398166, Minneapolis, Minnesota 55439.

Printed in the United States of America, North Mankato, Minnesota

052019
092019

THIS BOOK CONTAINS RECYCLED MATERIALS

Cover Photo: NSP-RF/Alamy
Interior Photos: NSP-RF/Alamy, 1; iStockphoto, 5 (top), 5 (bottom left), 5 (bottom right), 7, 9, 11, 14, 15, 20 (bottom right); Shutterstock Images, 6; Andrew J. Martinez/Science Source, 13; Mark N. Boulton/Science Source, 16–17; Leslie Newman & Andrew Flowers/Science Source, 18; Georgette Douwma/Science Source, 19; Andrew J. Martinez/Science Source, 20 (top); D. P. Wilson/FLPA/Science Source, 20 (bottom left)

Editor: Meg Gaertner
Series Designer: Jake Nordby

Library of Congress Control Number: 2018964520

Publisher's Cataloging-in-Publication Data
Names: Bassier, Emma, author.
Title: Starfish / by Emma Bassier.
Description: Minneapolis, Minnesota : Pop!, 2020 | Series: Ocean animals | Includes online resources and index.
Identifiers: ISBN 9781532163432 (lib. bdg.) | ISBN 9781644940167 (pbk.) | ISBN 9781532164873 (ebook)
Subjects: LCSH: Starfishes--Juvenile literature. | Invertebrates--Behavior--Juvenile literature. | Ocean animals--Juvenile literature.
Classification: DDC 593.93--dc23

Hello! My name is

Cody Koala

Pop open this book and you'll find QR codes like this one, loaded with information, so you can learn even more!

Scan this code* and others like it while you read, or visit the website below to make this book pop.

popbooksonline.com/starfish

*Scanning QR codes requires a web-enabled smart device with a QR code reader app and a camera.

Table of Contents

Chapter 1

Sea Star

Starfish, or sea stars, are ocean animals. Many starfish live in **coral reefs**. They live on the ocean floor. There are approximately 1,600 kinds of starfish.

Watch a video here!

Starfish can be different colors. Some starfish use **camouflage**. They blend in with their surroundings.

Other starfish are brightly colored to scare off **predators**.

Starfish have hard, rough skin. They have a central body with a mouth. They have many arms. Most starfish have five arms. One kind of starfish has 40 arms!

Starfish have one eye on the tip of each arm.

Starfish have thousands of tiny feet. Each foot is a sticky tube. The tube sticks to a surface. Then it lets go as the starfish moves along.

Despite their name, starfish are not fish.

arm
eye
feet
mouth

Chapter 2

Moving Stomach

Starfish eat clams, oysters, or mussels. These are shelled creatures that live on the ocean floor. Starfish use their feet to open the shells.

Learn more here!

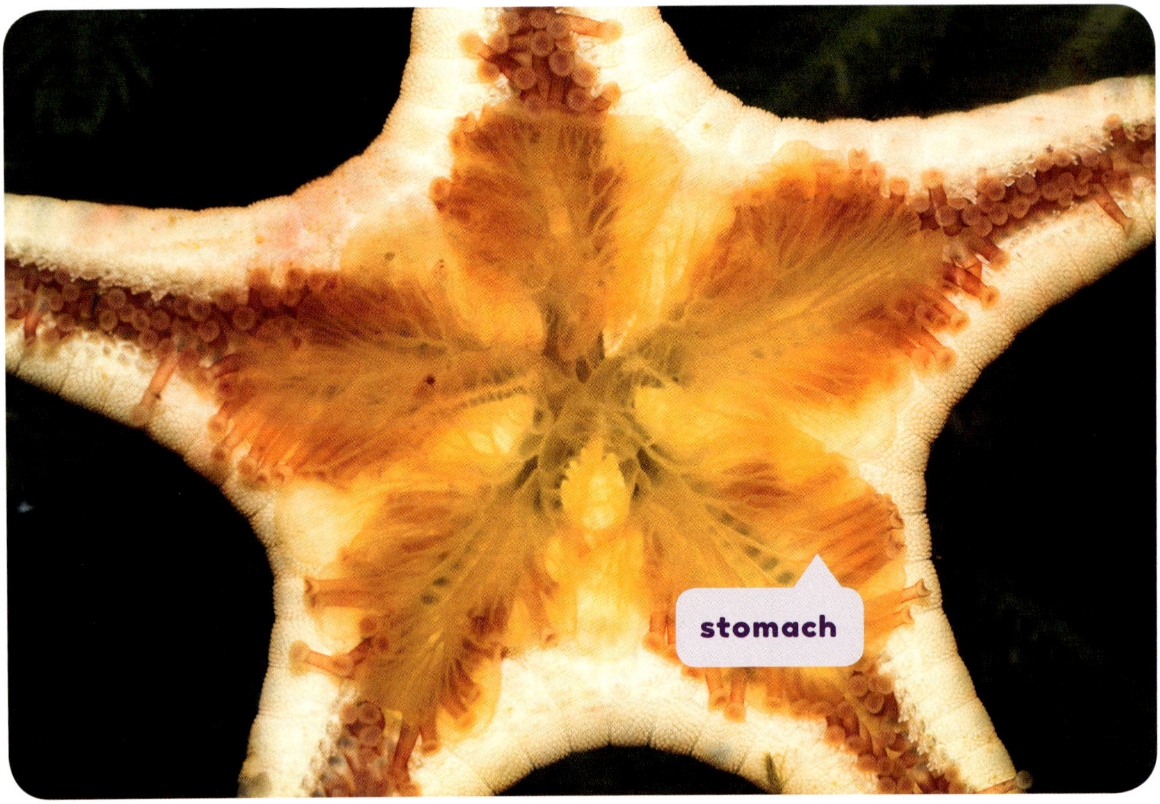

A starfish's **stomach** pops out of its mouth. The stomach wraps around **prey**.

It eats the prey. Then the stomach moves back inside the starfish's body.

Chapter 3

Regeneration

Starfish have a unique way of staying safe. They can grow new parts of their bodies. This process is called **regeneration**.

Learn more here!

A fish attacks. It bites off one of the starfish's arms. The starfish leaves the arm.

The starfish will survive. It will grow a new arm.

Chapter 4

Life Cycle

Many starfish release eggs into the ocean. The eggs drift with the water. Then they sink to the ocean floor. They grow into adult starfish.

Some starfish can live for 35 years.

Complete an activity here!

Making Connections

Text-to-Self

Starfish can regenerate body parts if they get hurt. What do you do when you get hurt?

Text-to-Text

Have you read other books about ocean animals? How are those animals similar to or different from starfish?

Text-to-World

Starfish have five or more arms. What other animals have many arms?

Glossary

camouflage – the ability to hide by blending in with one's environment.

coral reef – a shallow part of the ocean where corals live.

predator – an animal that hunts other animals.

prey – an animal that is hunted by other animals.

regeneration – the process of growing new body parts.

stomach – the part of an animal's body that breaks down food.

Index

Online Resources

popbooksonline.com

Thanks for reading this Cody Koala book!

Scan this code* and others like it in this book, or visit the website below to make this book pop!

popbooksonline.com/starfish

*Scanning QR codes requires a web-enabled smart device with a QR code reader app and a camera.